Table of Contents

The theory of General Relativity	2
Manifolds and tensors	6
The curvature	9
Lie derivatives and Killing fields	13
The character approach	15
The conformal approach	17
The thin-sandwich approach	19
York-Lichnerowicz conformal decomposition	20
The Hamiltonian and Momentum constraints	22
Hamiltonian formulation	26
Coordinate transformations	30
A regular metric in normal coordinates	33
Connection coefficient	34
The Bondi-Sachs	35
The Hyperbolicity concern	44
The Arnowitt-Deser-Misner equations	48
The Bona-Masso and NOR formulations	50
The Kidder-Scheel-Teukolsky family	53
References	59

The theory of General Relativity
The theory of General relativity postulated by Einstein at the end of 1915 and is the modern theory of gravitation. According to this theory, gravity is not a force as it used to be considered in Newtonian mechanics, rather a manifestation of the curvature of spacetime; a massive object produces a distortion in the geometry of spacetime around it, and in turn this distortion controls the movement of physical objects.

When Einstein introduced special relativity in 1905 it became clear that Newton's theory of gravity would have to be modified, The main reason for it was that Newton's theory implies that the gravitational interaction was transmitted between different bodies at infinite speed, in clear contradiction with one of the fundamental results of special relativity: "No physical interaction can travel faster than the speed of light"

It is interesting to note that Newton himself was not convinced with the existence of this action at a distance, but he considered that it was a necessary hypothesis to be used until a more adequate explanation of the nature of gravity was found.

The basics that guided Einstein in his quest towards general relativity were the principle of general covariance, which says that the laws of physics must take the same form for all observers, the principle of equivalence, which says that all objects fall with the same acceleration in a gravitational field regardless of their mass, and Mach's principle, formulated by Ernst Mach at the end of 19th century, which states that the local inertial properties of physical objects must be determined by the total distribution of matter in the universe.

The principle of general covariance led Einstein to ask for the equations of physics to be written in tensor form, the principle of equivalence led him to the conclusion that the natural way to describe gravity was identifying it with the geometry of spacetime, and Mach's principle led him to the idea that such geometry should be fixed by the distribution of mass and energy.

The discussion that follows will serve to present some of the basic concepts of general relativity, but it is certainly not intended to be a detailed introduction to this theory, it is simply too short for that.

Before going into general relativity it is important to discuss some of the basic concepts and results of special relativity. It was introduced by Einstein in 1905 as a way of reconciling Maxwell's electrodynamics with the Galilean principle of relativity itself, and second the empirical fact that the speed of light is the same for all inertial observers, a fact elevated by Einstein to the status of a physical law. The invariance of the speed of the light was established by the Michelson-Morley experiment in 1887, though it is unclear how much influence this experiment had in Einstein's development of special relativity.

Special relativity is in its essence a new kinematic framework on which we can do dynamics, that is, we can study the effects of forces on physical objects, so that accelerations are included in all times. Accelerating observers, or more to the point accelerating coordinate systems, can also be dealt with, though the mathematics become more involved. What makes special relativity "special" is the fact that it assumes the existence of global inertial frames, i.e. reference frames where Newton's first law holds: "Objects free of external forces remain in a state of uniform rectilinear motion".

Inertial frames play a crucial role in special relativity; In fact, one of the best known results from this theory are the Lorentz transformations that relate the coordinates in one inertial frame to those of another. The Lorentz transformations have a number of important consequences; The first of which can be easily derived by asking where the events that happen at $t=0$ according to the first frame of reference end up in the other frame of reference; meaning, events that happen at the same time $t=0$ in frame O and are thus simultaneous, happen at times that depend on their spatial

positions according to O' and are then not simultaneous, which implies, in particular, that the time order of events is not always fixed:
If two events are simultaneous in a given inertial frame O, then in frame O' moving with a non-zero speed 'v' with respect to O, one of the two events will happen at an earlier time than the other, while in a frame O" moving with speed '-v' with respect to O it will be the other way around.
The theory of special relativity was put on a more geometric foundation by Hermann Minkowski in 1908. Although Einstein initially perceived this as an unnecessary abstract way of rewriting the same theory, he later realized that Minkowski's contribution was in fact crucial for the transition from special relativity to general relativity. Minkowski realized that we could rewrite Einstein's second postulate about the invariance of the speed of light in geometric terms if we first defined the intervals between the two events; Einstein's second postulate then turns out to be equivalent to saying that the intervals between any two events is absolute, meaning, all inertial observers will find the same amount of interval, which means that we can define a concept of invariant distance between events, and once we have a measure of distance we can do the geometry; Notice that the ordinary three-dimensional Euclidean distance interval between

two events is not absolute, nor is the time interval, as can be easily seen from the Lorentz transformations; on the other hand, Minkowski's four-dimensional spacetime interval is absolute; in his own words "Henceforth space by itself, and time by itself, are doomed to fade away into mere shadows, and only a kind of union of the two will preserve an independent reality".

An important consequence of the Lorentz transformations is that the time order of events is in fact absolute for events with timelike fashion; this leads us to defining a notion of casuality in an invariant way: Events separated in a timelike fashion can be casually related, events separated in a spacelike way must be casually disconnected, as otherwise in some inertial frames the effect would be found to precede the cause. In particular, this implies that no physical interaction can travel faster than the speed of light as this would violate casuality - this is one of the reasons why in relativity nothing can travel faster than light.

Manifolds and tensors
Making the move from special to general relativity it is crucial to consider a few components of geometry. The basic knowledge of differentiation would be a big plus dealing with special relativity. It begins with a manifold and the act of differentiation on which. The spacetime of special

relativity is a four- manifold, while the sphere of a curved and closed, considered to be a two-manifold. Two particles located in the configuration space called a six-manifold and required six coordinates to provide the system with. There are some other properties common to manifolds not distant from one another. The rotational space in three dimensions over a rigid body is such; however, it has a complicated building block: Not only being closed, as it rotates an angle of 2pi, but also not having a trivial identity. A manifold has been defined as a space prepared to be covered by a collective one-to-one mapping. Lets consider an Euclidean which can be covered by the mapping from R^3 in a trivial fashion, on the other side over the surface of a sphere you might need two mappings from R^2. The image of a curve; being as functions from a point of the real line into the manifold, is called trajectory; The curve considered to be a function of points on the manifold. The change of coordinates does not change the curve, rather the function itself. The components of a vector includes derivative operators along the curve; Vectors at a given point can form a vector space; we can always define a vector as a linear mix of some other vectors. A general case of a choice vector is called coordinate basis, constituted of those vectors tanget to the coordinate lines.

Considering functions of vectors over their tangent; a linear, function of one vector, called a one-form. They will be denoted with a tilde and be written the act of a one-form over a vector. They also form a vector space with the same dimension as that of the manifold.

The definition of a basis for the space of one-forms, called dual, includes those such that, will give you identity matrix, when acting over the basis vectors. Using the linearity of the one-forms and defining the dual basis, the act of an arbitrary one-form on a basis vector can be generated, which is the sum of the components of a one-form times the components of vector; the result of which will be a real number independent of the coordinate system being used. One-forms and vectors are dual in one other regard: You might want to invert the definition of a vector as a real function of a one-form, i.e. a function that gives you back the real number q(v). Then you will be able to extend this argument and bring a real-valued functions of m one-forms and n vectors; that gives you a tensor; the components of which are the amounts of that applied to the parameters of the basis vectors and one-forms. An important one-form field is the gradient of a scalar function f from M into the real numbers. The indices for the components of the gradient are down, which tells you that it is a one-form not a vector; as it can be applied to an arbitrary vector to gain a real number, which is the directional derivative of f along the vector.

The symmetry properties of tensors with regard to the exchange of their arguments provides them with an important character. A tensor is symmetric if it remains the same with regard to exchange of any pair of arguments. In a similar fashion, a tensor called a completely anti symmetric if it changes sign with regard to exchange of any pair of arguments. These parts of an arbitrary tensor can be constructed by putting together all likely permutations of their arguments with appropriate signs.

The curvature

From the number of indices in $\Gamma\mu$ you could naively think that they are components of a tensor, but this is not the case. Looking at their definition, you can see that the connection coefficients map a vector v onto a new vector, so they are in fact coefficients of a tensors, one such tensor is a basis vector. This becomes more clear if you write the coefficients as $(\Gamma\alpha)\mu$, where the index α identifies the tensor and the indice μ...etc. etc. identifies the component of which. So, contrary to what is often said, the connection coefficients are in fact components of tensors; However this may not seem very useful since changing coordinates, you are also going to change the basis vectors and hence the tensors associated with the connection coefficients; because of these you can not expect the $\Gamma\mu$ or $(\Gamma\alpha)\mu$ for that matter to transform as components of a tensor, and indeed they don't. The transformation law for the connection coefficients is more complicated; Note that $(\Gamma\alpha)\mu$ is precisely what you would expect from the components of a tensor.

The transformation rules are in fact independent of any relation between the connection coefficients and the metric, i.e. it is applicable for any derivative operator. An important consequence of this is the following: Consider the connection coefficients associated with two different derivative operators and define $\Delta\mu:=\Gamma\mu-\Gamma°\mu$. It turns out that $\Delta\mu$ transforms as a tensor since cancels out; in another word, It is independent of the Γs.

$\Delta\mu$ is the component of a properly defined tensor. In a similar fashion, assume that you have a manifold with a given coordinate system and consider two different metric tensors $g\mu\nu$ and $g°\mu\nu$ defined on it.

The metric tensor in itself is not the most convenient way of describing a curved manifold as it can become quite non-trivial even in an Euclidean space considering curvi-linear coordinates. The right way to differentiate between flat and curved manifolds is by considering what happens to a vector as it is parallel transported around a closed circuit on the manifold. On a flat manifold, the vector does not change when this is done, while on a curved manifold it does. This can be clearly seen if you think about moving a vector pointing east. You move north following a meridian that is 90 degrees to the east of the first one until you get back to the equator.

In order to define a tensor that is associated with the curvature of a manifold you should consider the parallel transport of a vector along an infinitesimal closed circuit. If you take this closed circuit as one defined by the coordinate lines themselves, you can show that the change of the components of a vector Y as it is parallel transported along this circuit is given by: $\delta Y = R\beta\mu\nu \, dx\mu \, dx\nu \, \upsilon\beta$, where $R\beta\mu\nu$ are the components of the Riemann curvature tensor, which are given in terms of the Christoffel symbols as: $R\beta\mu\nu := \partial_\mu \Gamma\beta\mu - \Gamma\varrho\mu \, \Gamma\beta\nu - \Gamma\varrho\nu \, \Gamma\beta\mu$, Since a non-zero Riemann tensor indicates the failure of a global notion of parallelism you define a flat manifold as one for which the Riemann tensor vanishes, while a curved manifold will be one with a non-zero Riemann tensor. Notice that For Euclidean space in Cartesian coordinates the Riemann tensor vanishes. In fact, since the Riemann is a proper tensor, it must also vanish in any curvi-linear set of coordinates, the same true as of Minkowski spacetime.

The definition of the Riemann tensor also implies that for an arbitrary vector Y you should have: $\Delta_\mu \Delta_\nu Y^\alpha - \Delta_\nu \Delta_\mu = R^\alpha{}_{\beta\mu} Y^\beta$, this is known as the Ricci, and shows that the Riemann tensor also gives the commutator of the covariant derivatives of a vector. In other words, covariant derivatives of vectors commute on a flat manifold, but fail to commute on a curved one. Of course, this is the same thing as before as the commutator of covariant derivatives will tell you the difference when you parallel transport the vector to an infinitesimal close point following first one coordinate line and then another, or doing it in the opposite order.

The Riemann tensor has many symmetries, however the fully co-variant Riemann tensor is anti symmetric on the first and second pair of indices, and symmetric with respect to exchange of these two pairs. The symmetries of the Riemann tensor imply that in the end it has $n^2(n^2-1)/12$ independent components in n-dimensions, or twenty independent components in four dimensions. These symmetries also imply that the trace of the Riemann tensor; $R_{\beta\mu\nu}$, over its first and last pairs of indices vanishes. On the other hand, the trace over the first and third indices does not and is used to define the Ricci curvature; $R_{\mu\nu} := R_\mu{}^\lambda{}_\nu$;

The Ricci tensor is clearly symmetric in its two indices, and in four dimensions has 10 independent components: Note that in four dimensions having the Ricci tensor vanish does not mean that the manifold is flat.

Lie derivatives and Killing fields
Like vectors, tensors are defined at a specific point on a manifold, and although you can define tensor fields easily enough there is no natural way to compare tensors at different points in the manifold since they live on different spaces, which means, there is no natural fashion of the derivative of a tensor, since that requires you to compare the values of a tensor field at nearby points in the manifold. Given a metric, you can introduce a notion of derivative known as the covariant derivative. However, there is a more primitive way of doing that, namely, derivatives that do not depend upon the life of a metric; known as the Lie derivative. The congruence of integral curves of a smooth vector field on a region of a manifold, defines a mapping of the manifold into itself- you simply identify a given point with another point that lies on the same integral curve at a distance that defined by the change in the parameter defining the curve. You can then apply this mapping to drag tensors from the point to a point. It is easy to see how this works for a scalar

function: You point out that the dragged function is such that its amount on the next point is equal to the function of the first point. For a vector field you can define the dragging by looking at the curve associated with that vector field, and dragging the curve from a point to the next using the congruence associated with the vector field. The Lie derivative of a vector field is then defined in the following fashion: Evaluate the vector at the second point, drag it back to the first point, and take the difference with the original vector at the initial point in the limits when the distance comes close to zero. This indicates that the Lie derivatives depends upon the vector field used for the dragging. To find the components of the Lie derivative, we assign a new parameter let it be the integral curves of the vector field. If we drag an integral curve an infinitesimal distance along the vector field we will find that, the Lie derivative of u with respect to v is the commutator of v and u. Once you know how to drag vectors, you can also drag one-forms by asking for the amount of the dragged one-form when applied to the dragged vector to be equal to the amount of the original one-form applied to the original vector. In the same fashion, you can drag tensors of arbitrary rank. You can then define the Lie derivative of tensors in an analogous way to that of vectors. You might want to do the same thing for tensors of arbitrary rank, adding one more term for each index, with the adequate sign. There is an important property of Lie derivatives that helps considerably with their interpretation. Assume that we adapt one of our coordinates to the integral curves of the vector field v. In that case you see that the Lie derivative of a tensor T of arbitrary rank will simplifies and shows that the Lie derivative is a way to write partial derivatives along the direction of a given vector field in a fashion independent of the coordinates; a particular important application of which is related to the likely symmetries of a manifold that has a metric tensor defined upon it. You can say that the manifold has a specific symmetry if the metric is invariant under Lie dragging with regard to some vector field, that from the expression for the Lie derivative of a tensor you find that it implies, if given a metric $g_{\mu\nu}$ there lives a vector field ξ that satisfies the following equation; $\xi^\mu \partial_\mu g_{\alpha\beta} + g_{\alpha\mu} \partial_\beta \xi^\mu + g_{\mu\beta} \partial_\alpha \xi^\mu = 0$, then ξ is called a killing field; is easy to see the existence of which implies a symmetry of the manifold.

The character approach

It originates from the idea of using a foliation of spacetime based not on spacelike, rather on null hypersurfaces. The use of null hypersurfaces is very attractive if you are interested in extracting gravitational wave information from a macro system. Working with spatial hypersurfaces, this extraction typically requires the use of perturbative expansions around some type of a Schwarzschild; on the other hand, there is a completely rigorous description of gravitational waves on null hypersurfaces even in the non-linear context, that is the main motivation using the character approach, yet it has another important advantage. In a more practical term, we can only evolve a finite region of spacetime, one after another, so when using spatial hypersurfaces you will be forced to introduce some artificial/imaginary boundary condition at a finite distance; it then turns out that you might have to add a second boundary surface in order to have a non-trivial future domain of dependence. Typical choices are to use either a second null hypersurface or a central timelike world-tube that in some cases is collapsed to a single world-line.

In the character approach, we consider a foliation of null hypersurfaces corresponding to the level sets of a coordinate function U; on each hypersurface, extra radial coordinate λ being used, in terms of the specific form of the equations depends on the choice of the form of the metric; a very common choice would be to use the Bondi-Sachs coordinate system, which in the general three-dimensional case corresponds to a spacetime metric of the form, in which, the radial coordinate will be used instead of λ, thereafter, you will find hypersurface equations, involving just derivatives inside the hypersurface for the metric functions and evolution equations involving derivatives with respect to the null coordinate U for the metric functions.

One important advantage of this approach is the fact that there are no elliptic constraints on the data, so the initial data is free. In addition to that, there are no second derivatives in time; along the direction U, so there are fewer variables than a 3+1 approach. Plus, null infinity can be brought to a finite distance in coordinate space, that no artificial/imaginary boundary conditions are required; a lot of work has been devoted to developing character codes in spherical and axial symmetry, and today there are also well-developed three-dimensional codes that have been used to study, for instance, scattering of waves by a black hole; wave packets get compressed as seen in coordinate space as they move outward, which causes a gradual reduction in resolution acts as a change in the refraction index that in turn will cause waves to be back-scattered.

The conformal approach
In this approach, spacetime is foliated into spacelike hypersurfaces that reach null infinity. The

metric of this hypersurface is clearly positive/definite for all x, i.e. it is spatial everywhere, If you attempt to write down the Einstein field equations for the conformal metric in a straightforward fashion, it turns out that they are singular at places where the conformal factor is zero; however, a regular form of the conformal field equations has been derived by Friedrich, the variables involved are the connection coefficients, the trace and trace-free parts of the Riemann curvature tensor; the Weyl tensor, plus the conformal factor and its derivatives.

Friedrich's system of equations can also be shown to be symmetric hyperbolic. In the conformal formulation you are evolving the conformal factor as an independent function, hence, the position of the boundary of spacetime at null infinity is not known a priori, you will then need to extend the physical initial data in a smooth fashion and evolve the dynamical variables. This has been one important advantage, namely that it is possible to put an arbitrary boundary condition at the outer boundary of the computational region without affecting the physical spacetime.

The conformal formulation would seem to be an ideal solution to the weaknesses of both the standard 3+1 approach and the characteristic formulation. Being based on spatial hypersurfaces, it does not have to deal with the problem of caustics associated with the character formulation; at the same time, by reaching close to null infinity, it allows extraction of gravitational radiation and other quantities such as total mass/momentum. The main concern though, faced by the conformal formulation being related to the problem of constructing hyperboloidal initial data; being based on spatial hypersurfaces, it will have to solve many of the same problems that standard 3+1 formulations are currently faced with, i.e. the choice of a good gauge and the stability of the evolutions against constraint violation.

The thin-sandwich approach
Out of the ten Einstein field equations, six contain time derivatives and therefore represent the true evolution equations of the spacetime geometry. The remaining four equations are constraints that must be satisfied at all times. These are the Hamiltonian and momentum constraints, the existence of which implies that it is not possible in general to choose arbitrarily all twelve dynamical quantities as initial data.
From the beginning, If the initial data is not being chosen in a fashion that the constraints are satisfied, you won't be solving Einstein's equations, which means before starting an evolution, it is necessary to first solve the initial data problem and obtain adequate amounts representing the physical

situation you are interested in.

The constraints form a system of four coupled partial differential equations of elliptic type, and in general they are difficult to solve; however, there are several well-known procedures to solve them in specific circumstances. More recently, the so called Conformal thin-sandwich approach has become more and more popular, solving the constraints, as it allows for a clearer interpretation of the freely specifiable data.

York-Lichnerowicz conformal decomposition

Trying to solve the constraint equations, one is immediately faced with the problem of having four differential equations for the twelve degrees of freedom associated with the spatial metric and extrinsic curvature. The first question that must be answered is which of the 12 quantities will be taken as free data, and which will be solved for using the constraints. Except in very simple cases like that of the linearized theory, there is no natural way of identifying which are the true dynamical components.

You should therefore develop some procedure that chooses eight components as free data and allows to solve for the remaining in a clear way. The most common procedure for doing that is known as the York-Lichnerowicz conformal decomposition, starts from a conformal transformation of the three-metric, where the conformal metric is considered as given.

In terms of the conformal metric, the Hamiltonian constraint takes the Laplace operator and Ricci scalar associated with the conformal metric. The extrinsic curvature is also separated into its trace and its tracefree part. The Hamiltonian constraints then will be transformed into an elliptic equation for the conformal factor ψ, and solving it will clearly allow you to reconstruct the physical metric from a given conformal metric.

In order to transform the momentum constraints into equations, you can state that any symmetric-tracefree tensor can be split into a symmetric, traceless, transverse tensor, and a vector and an operator L.

Notice that the operator L can be defined using any metric tensor. Two natural choices present themselves at this point for the decomposition of momentum constraints, thereof the conformal Ricci tensor appears in the last expression when you commute covariant derivatives.

Even though it is a simple task to find a symmetric/trace-free tensor, it is quite a different matter to construct a transverse tensor. In order to construct such a tensor you need to start from an arbitrary symmetric/trace-free tensor that is not necessarily transverse. It is common to define the energy and momentum densities, the weight of the conformal factor in the definition of them is chosen in order to eliminate factors of ψ from the matter terms in the momentum constraints and thus decouple them more easily from the Hamiltonian constraint.

The weight of ψ in the definition of the energy/momentum density is then fixed for consistency reasons; this would be the most common path toward writing the constraints in the York-Lichnerowicz approach. To simplify the approach considerably you can choose K constant, corresponding to a constant mean curvature spatial hypersurface, in which case the momentum constraints decouple completely from the Hamiltonian. The approach simplifies even more if apart from taking K constant, one also takes the conformal metric to be the one corresponding to flat space.

The Hamiltonian and Momentum constraints
In a coordinate system adapted to the foliation, the Hamiltonian and momentum constraints take the final form; It is important to notice that the constraints not only do not involve time derivatives, but they are also completely independent of the gauge functions. This indicates that the constraints are relations that refer to a given hypersurface. Having a set of constraint equations is not a feature of general relativity

alone. In electrodynamics you have the Maxwell equations which in three-dimensional vector calculus notation, and in Gaussian units, take the form of the product of electric and magnetic fields respectively. It will give you two equations involving the divergence of the electric and magnetic fields which do not involve time derivatives, so they are in fact constraints, just as in general relativity.

The remaining two Maxwell equations; or rather six as to their vector-valued equations, are the true evolution equations for electrodynamics. The existence of constraints implies, in particular, that in 3+1 formulation it won't be possible to specify arbitrarily all twelve dynamical quantities as initial conditions.

The initial data must already satisfy the constraints, otherwise you won't be solving Einstein's equations. The Hamiltonian and momentum constraints give you four of the ten independent Einstein field equations, and they do not correspond to the evolution equations of the gravitational field, rather to relations between the dynamical variables that must be satisfied at different points in time. In order to find these equations you need the projection onto the hypersurfaces of the Riemann tensor contracted twice with the normal vector. This will give you the last six independent components of

Riemann/extrinsic curvature, which together with the equations for the evolution of the spatial metric, they will let you write down the field equations for general relativity as a Cauchy. The first thing to notice is that these relations do involve the lapse function, also they make reference to the Lie derivative of the extrinsic curvature along the normal direction, which clearly corresponds to evolution in time; or expanding the Lie derivative along the shift vector; These relations give us the dynamical evolution of the six independent components of the extrinsic curvature. Together with equations for the evolution of the spatial metric they will let us to write down the field equations for general relativity as a Cauchy problem; It is important to notice that we do not have evolution equations for the gauge quantities that represent your coordinate freedom. The evolution equations are known as the Arnowitt-Deser-Misner; however, these equations as represented above, are in fact not in the form from which stemmed; however, they are a non-trivial rewriting due to York. The difference between the Arnowitt-Deser-Misner and York evolution equations can be traced back to the fact that the version of Arnowitt-Deser-Misner comes from the field equations written in terms of Einstein tensor, whereas the version of York was instead derived from the field equations written in terms of the Ricci tensor.

Both sets of evolution equations in terms of their physical solution are equivalent since they only differ by the addition of a term proportional to the Hamiltonian constraint; as for the former, the original Arnowitt-Deser-Misner variables are the spatial metric and its canonical conjugate momentum coming from the Hamiltonian formulation of general relativity; Note that the goal of Arnowitt-Deser-Misner is to write a Hamiltonian formulation for general relativity that could be used as a basis for quantum gravity and not a system of evolution equations for dynamical simulations; also should be noted that the different evolution equations in terms of their mathematical application are not equivalent; due to two basic reasons:

First, the space of solutions to the evolution equations is different in both cases, and only coincides for physical solutions, that is, those that satisfy the constraints. Of course, we could always argue that since in the end we are only interested in physical solutions, which is true only if we can solve the equations exactly; The crucial factor here might be: If you move slightly off the constraint hypersurface, does the subsequent evolution remain close to it, or does it diverge rapidly away from it?

Secondly, the Hamiltonian constraint has second

derivatives of the spatial metric; hidden inside the Ricci scalar, then by adding a multiple of it to the evolution equations you are in fact altering the structure of the differential equations.

The 3+1 evolution equations are highly non-unique since you can always add to them arbitrary multiples of the constraints. The different systems of evolution equations will still coincide in the physical solutions, but might differ in their mathematical properties, and in particular, by the fashion in which they react to small violations of the constraints; The other consideration about them has to do with the propagation of the constraints: Do the constraints remain satisfied throughout the course of the evolution?! What if we use York's evolution equation?!

Hamiltonian formulation

The field theoretical formulation of general relativity starts from the Hilbert Lagrangian which has been introduced: L=R, with R the Ricci scalar of the spacetime. The Lagrangian formulation of a field theory takes a covariant approach. First, the Lagrangian itself should be a scalar function, and also the field equations derived from the variational principle come out in fully covariant form; a different approach would be to take instead a Hamiltonian formulation of the theory; it requires a clear distinction to be made between space and time, so it is therefore not covariant.

In field theories other than general relativity, and particularly when working on a flat spacetime background, there is already a natural way in which space and time can be split. In general relativity, on the other hand, no such natural splitting exists; However, you can not interpret the time function directly as a measure of the proper time of any given observer.

The first step in a Hamiltonian formulation is to identify the configuration variables that describe the state of the field at any given time. To this end, you will have to choose the spatial metric variables, together with the lapse and the co-variant shift vector. Then you are going to rewrite the Hilbert Lagrangian in terms of these quantities and their derivatives. Notice that, from the definition of the Einstein tensor, in its second term, you will have the Ricci identity that relates the commutator of covariant derivatives to the Riemann tensor; also Note that You have already been provided with some terms of the extrinsic curvature even though no projection operator is present. You can readily verify that the contractions in these expressions guarantee that the result follows.

The last term of the Ricci scalar takes a total divergence, and since in the end you are only interested in an integral of the Lagrangian over a given volume, this term can be transformed into an integral over the boundary of the volume and can therefore be ignored.

Notice that L has a similar infrastructure to that of the Hamiltonian constraint, with the sign of the quadratic terms in the extrinsic curvature reversed.

After ignoring a total divergence, with H and Mi, but without the matter contributions, that would arise when you add the Hamiltonian density for the matter; the total Hamiltonian will be created, The variation of which with respect to lapse and shift function, immediately yields the Hamiltonian and momentum constraints for vacuum, H=0 and Mi=0. In other words, the lapse and shift behave as Lagrange multipliers.

Notice when using the hamilton equations to derive the 3+1 evolution equations, we often take the lapse function as an independent quantity to be kept constant throughout the course of variation; however, you might want to adopt a different point of view and assume that the independent gauge function is not the lapse.

The Hamiltonian then reduces to a new equation which includes a D^2-flat term that is the flat space Laplacian. In the particular case of time-symmetric initial data, corresponding to $K_{i,j}=0$ at $t=0$, the momentum constraints are trivially satisfied. If, moreover, we assume that we are in vacuum, the Hamiltonian reduces further to D^2flat $\psi=0$, which is the standard Laplace.

The conformal transverse decomposition results in somewhat simpler equations, and because of this it has been used more in practice. The existence of two different methods of splitting the traceless extrinsic curvature might suggest that perhaps neither is optimal. The root of the problem lies in the fact that the conformal rescaling and tensor splitting do not commute.

More explicitly, from the identity S_{ij}; the symmetric tracefree tensor, we learn that the natural conformal transformation for that would have a ψ-10 part; However, the natural conformal transformation of the longitudinal part is instead the mismatched powers of ψ in these transformation rules are at the root of the non-commutativity of conformal transformation and tensor splitting.

The main argument is to split the tracelss extrinsic curvature A_{ij} as $A_{ij}=A^*_{ij}+1/\sigma(LW)_{ij}$, with σ a positive definite scalar. This is certainly more consistent and elegant splitting than the previous cases, produces yet another form of the momentum constraints that now depends upon a new scalar weight that has to be chosen as free data; the immediate question then will be; what physical meaning one can give to this weight function ?!

The York-Lichnerowicz conformal decomposition described, gives you a clear way to find solutions to the constraint equations, i.e. initial data, starting from some free data given in the form of a conformal metric, a symmetric/trace-free tensor, the trace of the extrinsic curvature K, and the energy/momentum densities. However, there is in general no clear insight as to how to choose the symmetric/trace-free tensor to represent a particular dynamical situation, and worse, the different possible decompositions will give you different initial data starting from identical free data. Arguably, the weighted decomposition of Pfeiffer and York is the best approach, but in that case you will have to specify yet another free function in the form of the scalar weight, for which there is no clear physical interpretation.

Coordinate transformations

Up until now we have assumed that we are given a specific coordinate system and its associated coordinates basis. However, we should also consider what will happen when we transform to a

different set of coordinates. This is important as coordinates are in fact arbitrary labels for points in a manifold, and you might want to choose different coordinates under different circumstances. For instance, in the particular case of special relativity, you wish to use the coordinates associated with a given inertial frame, or those associated with a different inertial frame and related to the first ones via Lorentz. You might also wish to consider arbitrary changes of coordinates. It is easy to see that under a change of coordinates the components of the displacement vector transform to the Jacobian matrix. In the particular case of special relativity, the Jacobian identity for the Lorentz transformations is given, but you can consider more general/non-linear, changes of coordinates. An important property of a change of coordinates is that, in the region of interest, the transformation should be one on one, as otherwise the new coordinates wouldn't be useful, which implies that the Jacobian is always invertible in that region. From the definition of the components of a vector it is easy to see that they transform just as the displacement vector.
When you transform the coordinates, you have clearly also changed your coordinate basis, as the new basis must refer to the new coordinates. From the fact that a vector as a geometric object is invariant under coordinate transformations you

can derive the transformation law for the basis vectors themselves, where now the Jacobian is the of the inverse transformation; in a similar fashion you can find the transformation laws for the components of a one-form and the one-forms that form the dual basis. As to the form of the transformation laws, you often refer to lower indices as co-variant since they transform like the basis vectors; with the inverse Jacobian. As an example of these transformation laws, lets consider the gradient of a scalar function f. We have already mentioned that the gradient is a one-form and not a vector. You can see this directly by considering the transformation of the components of the gradient that transform as those of a one-form and not as those of a vector.

You can easily extend the transformation laws to tensors of arbitrary rank, the rule is to use one Jacobian factor to each index, using either the direct or inverse Jacobian depending on the position of the index. In special relativity, the change of coordinates is often given by the Lorentz transformations between inertial frames, which are linear in the arguments. However, you will need to consider more complex non-linear coordinate transformations even on an Euclidean space.

A regular metric in normal coordinates

Let (M, g) be the four-dimensional space-time. Suppose (M, g) contains a simply convex normal neighborhood Γ with a regular time-like geodesic $c(\tau)$, where $y\alpha$ is its unit, time-like tangent vector and τ the proper time. Since Γ is a convex normal neighborhood and $c(\tau)$ is contained in Γ, a regular metric $g_{\alpha\beta}$
can be found along $c(\tau)$. Let $y\alpha$ be a set of four-dimensional arbitrary coordinates along $c(\tau)$ such that the metric $g_{\alpha\beta}$ is regular and can be expanded along $c(\tau)$ where the coefficient are evaluated along $c(\tau)$. The timelike base vector of the coordinate set e0 is tangent to $c(\tau)$, i.e. $e0(\tau) = \partial/\partial\tau$. On any point of the geodesic $c(\tau)$, you send out space-like geodesics $b(\tau,n,l)$ which are parametrized by an affine parameter l and point into the direction $n = n_i\, e_i(\tau)$, i.e. $n0 = 0$ and $\partial/\partial\ell\,|\,c(\tau)$. The parameter l of the geodesics $b(\tau,n,l)$ is defined to be zero along $c(\tau)$.

Note that the three-dimensional coordinates y_i are Riemann normal coordinates for every amount of τ on $c(\tau)$. constructing the coordinates in this fashion assures the metric along $c(\tau)$ is that of a Minkowski, because the base vectors are orthonormal along $c(\tau)$. In addition, it implies the first order partial derivatives vanish along $c(\tau)$, because of the parallel transport equation of n_i along $c(\tau)$ and the geodesic equations of $c(\tau)$ was first derived by Misner and the corresponding metric are spacetime invariants, i.e. knowing the independent Riemann normal components along $c(\tau)$ allows you to construct uniquely a metric up to quadratic terms of a power series expansion with respect to normal coordinates along timelike geodesic in a sufficiently small neighborhood of the geodesic.

Connection coefficient

It is clear at this point that when you consider the change in a tensor field as you move around the manifold, you should take into account not only the change in the components of the tensor, but also the fact that the basis in which those components are calculated might change from one point to another. The basic problem here is that, as mentioned before, there is in general no natural way you give the manifold some more structure. You can in fact start in a very abstract way by defining a derivative operator Δ. You get one extra index down because you can take derivatives along any direction, an operation you represent by $\Delta\alpha$. This operator must have a series of properties in order to be identified as a derivative: It must be linear, it must follow the Leibnitz rule for the derivative of a product, it must reduce to the standard partial derivative for scalar functions, and it must be symmetric in the sense that for a scalar function f you get $\Delta\alpha\Delta\beta\ f=\Delta\beta\Delta\alpha\ f$. Once you have an operator Δ you can consider, for instance, the derivative of a vector with regard to a given coordinate.

Notice that the derivative of a vector is more than just the derivative of its components. You should also take into account the change in the basis vector themselves. Now, if you choose a fixed direction the derivative followed by must in itself be also a vector, since it represents the change in the basis vector along that direction, which means, it can be expressed as a linear combination of the basis vectors themselves. Next you introduce $\Gamma\mu/\alpha\beta$ to denote the coefficients of such linear combination. This is known as the connection coefficients, as it allows you to map vectors at different points in order to take their derivatives.

The Bondi-Sachs

The work of Bondi-Sachs contributed to a deep understanding of gravitational waves at distance from source. They applied a coordinate map to the metric that was meant for null geodesics creating null cones. "Retarded Time" υ, plus the spherical angles θ and φ.

They showed that holding an observer at infinitely far distances from an isolated gravitational object, the system loses mass/energy only through gravitational radiation. An observer at a null infinity can only keep the track of an event in the future or past of a source that radiates. If the concern is the study of the procedure itself, then you might want to consider what is happening at the source. If you place an observer on a timelike world tube of finite spacelike radius; where outgoing null geodesics are attached, you will be able to form the basis of many numerical codes that are being used to solve the Einstein equations. When the radius becomes zero, the world tube degenerates into a line, i.e. into a curve of an observer tracking the out-going null cones plus the origin of the coordinate set.

Isaacson integrated the Einstein equations for the Bondi when the radius is zero. Then he chose the boundary conditions for the Bondi metric such that it approaches flat space amounts at the origin. His work motivated Gomez and Siebel to solve the vacuum Einstein equations for the Bondi in axial symmetry. The boundary conditions for the Bondi metric included the lowest order curvature. Here we are dealing with the boundary conditions of the Bondi-sachs; a vertex of a null cone would be your focal point where null geodesics making the cone converge from. In a mathematical term, you might not be able to differentiate the null cone at its vertex; in particular the curvature tensor can not be known of. Therefore, the boundary conditions in Bondi-Sachs coordinates are not known at the vertex.

Finding the appropriate boundary conditions on a null cone at the vertex, required:

1) The vertices of the null cones are traced by a world line of an observer to be a time like;

2) The geodesics are contained in a normal neighborhood, such that it has the property that two or more points in that neighborhood can be connected by a unique geodesic;

These prerequisites are important trying to assign the boundary condition of the Bondi-Sachs. Tn initial observer's metric moving along a time-like geodesics, you might want to use coordinate transformations to find a Bondi-sachs metric, that will not be regular because the angular base vectors are not defined there. Based on rectangular coordinate set, you will then define a metric on a null cone that is subsequently transformed into a Bondi-Sachs, through a change in the radial coordinate turning it over into an areal distance coordinate; this procedure helps split the Einstein equation into a set of hypersurface equations and evolution equations. You might want to start with a regular metric along the time like geodesic in another coordinate set and finding the boundary conditions for a Bondi-Sachs via a proper coordinate transform.

Here our aim is to present a complete derivation of the boundary conditions of a Bondi-Sachs metric at the vertex from a regular metric of an inertial observer along a timelike geodesic. That said, we are provided by additional steps in the derivation and expansion of the results. You must agree that there is no need of a system of a bi-tensor theory to find a null metric with an affine parameter as radial coordinate. To further analyze the vertex problem, the first point you can make would be what questions are going to be raised to help formulating boundary conditions of a Bondi-Sachs.

Notice that regularity issues at the origin of a curvi-linear coordinate system are also of some degree of importance in the 3+1 formulation of General Relativity, The first in-depth analysis of which done by Bardeen-Piran/1983. After defining their notation, they derived a limiting behavior of a Bondi-sachs metric from a regular normal coordinate set.

Recently, there has been an increased interest in formulations for numerical relativity based on conformal compactification in which the calculational grid extends to future null infinity where the gravitational radiation amplitude can be read off unambiguously, with at most numerical errors, and where no dynamical boundary conditions are necessary. In principle this can be done with Cauchy characteristic matching methods, but these have not been implemented successfully with non-trivial dynamics. Cauchy characteristic extraction methods can extrapolate from the outer boundary of a conventional Cauchy code to determine waveforms, but do not eliminate errors in the Cauchy development deriving from inexact boundary conditions at a finite radius. The characteristic methods are reviewed by J. Winicour. Our focus here is the initial value problem on hyperboloidal spacelike hypersurfaces, and specifically the case of constant mean extrinsic curvature (CMC) hypersurfaces. The vanishing of the conformal factor Ω at accounts for the singular behavior of the physical spacetime metric at in compactified coordinates. The conformal geometry, in which future null infinity is an ingoing null hypersurface, is regular in a neighborhood of future null infinity, and a CMC hypersurface of the physical spacetime is spacelike in the conformal spacetime out to and including its intersection with future null infinity, a 2-surface with spherical topology. Regularity conditions relating the 2D extrinsic curvature of as embedded in the CMC hypersurface to the conformal extrinsic curvature need to be satisfied at future null infinity, but when imposed in the initial data are automatically preserved by the evolution equations. The constraint equations, given suitable gauge conditions, determine the leading behavior of the conformal factor and the conformal extrinsic curvature of the hypersurface in the neighborhood of future null infinity, in terms of asymptotic gravitational wave amplitudes. If the physical mean extrinsic curvature is not too large, the dynamics of the sources takes place where the CMC hypersurfaces are not that different from the hypersurfaces of conventional 3+1 methods.

The conformal momentum constraint equations have the same form as they do on maximal hypersurfaces, and with conformal flatness recognize a class of analytic solutions which are slight generalizations of the well-known Bowen-York solutions often used for single or multiple initial data on maximal hypersurfaces. The Hamiltonian constraint provides an elliptic equation for the conformal factor which is degenerate at future null infinity. Despite this degeneracy, Ref. [3] obtained numerical solutions without much difficulty using the standard spectral elliptic solver of the Caltech-Cornell-CITA SpEC code [14,15].

The degeneracy constrains the leading terms in the expansion of Ω about future null infinity. An essential part of the physical interpretation of these solutions is knowing precisely the total energy, linear momentum, and angular momentum of the system as coded in the asymptotic behavior of the spacetime metric at future null infinity. The standard Arnowitt-Deser-Misner formulas for these quantities only apply on asymptotically flat slices at spatial infinity. The Bondi-Sachs energy and momentum are the relevant quantities for CMC hypersurfaces. Ref. [3] did not fully address this issue, making only some rather crude estimates of the total energy, momentum, and angular momentum with limited validity and for the most part with uncertain errors. While there is an extensive literature dealing with the problem of extracting these global physical quantities near or at future null infinity, part of which deals specifically with CMC hypersurfaces, we see practical difficulties in implementing many of these procedures. Some of them require the choice of a reference spatial metric and extrinsic curvature. Determining the reference quantities with the precision necessary to obtain unambiguous results requires considerable effort, in general. Formulas based on the asymptotic behavior of the Weyl tensor can be useful.

We first consider the problem of calculating the Bondi energy and momentum in general asymptotically flat space-times foliated by CMC hypersurfaces, then we analyze the asymptotic geometry using Gaussian normal spatial coordinates tied to the 2-surface. This gauge choice allows a relatively simple characterization of the asymptotic behavior of the spacetime metric and extrinsic curvature on a CMC hypersurface based on asymptotic solutions of the constraint equations. Constructing the asymptotic coordinate transformation from the CMC-based coordinates to Bondi-Sachs null coordinates gives us the Bondi-Sachs mass aspect in terms of geometric objects defined on a single CMC hypersurface. The monopole and dipole moments of the Bondi-Sachs mass aspect are the Bondi-Sachs energy and linear momentum, respectively. In the simple case of conformally flat initial data, the mass aspect is just a sum of contributions from the asymptotic expansions of the conformal factor and conformal extrinsic curvature.

The Bowen-York solutions for the conformal extrinsic curvature contain parameters such as boost and spin vectors, which on maximal hypersurfaces are simply related to the ADM momentum and angular momentum. On CMC hypersurfaces, the relationship of the Bowen-York parameters to physical momenta and energies is more complicated, but we are able to obtain analytic expressions for the direct contributions of the conformal extrinsic curvature to the Bondi-Sachs energy and momentum in terms of the Bowen-York parameters. Unlike on maximal hypersurfaces, on conformally flat CMC hypersurfaces the coordinate displacement of the black hole from the center of the coordinate sphere representing enters in a non-trivial way. The conformal factor contributions to the Bondi-Sachs energy and momentum are extracted from the numerical solution of the Hamiltonian constraint equation. We also derive an analytic expression for the total angular momentum of the system which only depends on the Bowen-York parameters, without any contribution from the conformal factor. We discuss the physical interpretation of some representative examples of Bowen-York initial data on CMC hypersurfaces similar to those of Ref. [3], and show how to construct initial data approximating a circular-orbit binary black hole system.

The Wald sign convention for the extrinsic curvature is adopted, so that the mean extrinsic curvature is positive for hypersurfaces with diverging future-directed normals, the hypersurfaces extending to future null infinity rather than to spatial infinity.

We assume vacuum in a neighborhood of future null infinity; we use the lapse and shift which preserve the CMC hypersurface condition and the Gaussian normal spatial coordinate condition to facilitate the asymptotic transformation to Bondi-Sachs coordinates. The final expression for the mass aspect depends only on the geometric properties of a single CMC hypersurface.

The Hyperbolicity concern

Lets look at a first order set of evolution such as the following equation; $\partial t\, \upsilon + M_i\, \partial i\, \upsilon = \upsilon's$, where M_i are matrices of $n*n$; with i walks on the spatial dimensions, and the source term $\upsilon's$. There are various options when you are concerned with hyperbolicity. This concept, i.e. hyperbolicity is mostly associated with set of equations dealing with evolution that should be treated as an extension of the simple wave. These are well-posed sets, at the same time, they hold onto a finite speed of propagation. You can begin with the definition of hyperbolicity built on the properties of M_i. There is an arbitrary unit vector n_i, through which the marix $P(n_i) := M_i\, n_i$, will be built. The P matrix has real eigenvalues and a complete set of eigenvectors for all n_i; when P has real eigenvalues for all n_i but does not constitute a complete set of eigenvectors, the system is still hyperbolic. For a hyperbolic system with a complete set of eigenvectors you can all the time find a positive/definite Hermitian; $H(n_i)$ such that; $HP - P^T H^T = HP - P^T H = 0$, where the super index T is the transposed matrix; the new matrix HP is symmetric.

If the system is strongly hyperbolic (a complete set of eigenvectors e_a such that; $P e_a = \lambda_a\, e_a$, index a walks on the dimensions of the υ) with λ_a the corresponding eigenvalues. R as the matrix of column eigenvectors, can be inverted since all the eigenvectors are linearly independent. HP is symmetric by the notion that $R^{-1} P R = \Lambda$, You can then conclude, Λ is diagonal so that Λ to the power of T is equal to Λ, which implies $(R^{-1})^T \Lambda R^{-1}$ is symmetric, the eigenvectors "ea" are defined up to an arbitrary scale factor. The system of equations is symmetric hyperbolic if all the M_i are symmetric, in other words, if the symmetrizer is independent of unit vector.

You can show that hyperbolicity and the existence of a conserved energy norm are equivalent, so instead of analyzing the P matrix you might want to look directly after the existence of a conserved energy to show that a system is hyperbolic. Note that for symmetric hyperbolic systems the energy norm will be independent of the unit vector, also note that, as was the case with the eigenvectors, the eigenfields are defined up to an arbitrary scale factor.

Lets consider case of a single dimension x. Multiplying the equation; $\partial_t \upsilon + M^i \partial_i \upsilon = \upsilon$'s with R^{-1} on the left of her, you will find that; $\partial_t \omega + \Lambda \partial_x \omega = 0$, with $\upsilon = R \omega$ or $\omega = R^{-1} \upsilon$; so to decouple the evolution equations of the eigenfields. Now that we have a set of independent equations, each with a speed of propagation given by λa; which is the mathematical notion that associates a hyperbolic set with having independent wave fronts propagating at finite speeds; it's worth to note that in the multi-dimensional sets, the full system will often not decouple even for symmetric hyperbolics, as to the fact that the eigenfunctions will depend upon the unit vector ni.

You can also use the eigenfunctions to study the hyperboloicity of a system; the argument here would be to construct a complete set of linearly independent eigenfunctions ωa that evolve via simple equations starting from the original variables υa. If this should be possible then the set will be strongly hyperbolic.

The initial assumption up until now has been that the marices Mi have constant coefficients, and also the source term s(υ) vanishes. In the more general case when s(υ) is not equal to zero and Mi=M(t,x,υ) you can still define hyperbolicity in the same fashion by linearizing around a background solution υ(t,x) and considering the local form of the matrices Mi.

The main difference is that now you can only show that solutions exist locally in time, as after a finite time singularities in the solution may develop; also, the energy norm does not remain constant in time but rather grows at a rate that can be bounded independently of the initial data. A particularly important case is that of quasi-linear set of equations where you have two different group of variables u and v such that derivatives in both

space and time of the u's can always be expressed as combinations of v's, and the v's evolve through equations of the form $\partial_t v + M_i(u) \partial_i v = s(u,v)$, with the matrices M_i functions only of the u's. In such a case you can bring the u's freely in and out of derivatives in the evolution equations of the v without changing the principal section by replacing all derivatives of u's in terms of v's; In the Einstein field equation you will see the same property, with the u's representing the metric coefficients; lapse, shift and spatial metric and the v's representing both components of the extrinsic curvature and spatial derivatives of the metric. The simplest example of a strongly hyperbolic equation is the one-dimensional advection equation itself; $\partial_t u + v \partial_x u = 0$, with v a real constant. The solution of this equation propagates the initial data with a speed v without changing its original profile. In other words, if $u(t=0,x)=f(x)$, then
$u(t,x) = f(x-vt)$; Advection terms like this appear all the time in the 3+1 evolution equations. The next interesting example is the one-dimensional wave equation; $\partial^2_t \varphi - v^2 \Delta^2 \varphi = 0$, with v the wave speed. This equation can be written in first order form by defining; $\Pi := \partial_t \varphi$, $\Psi_i := v \partial_i \varphi$; the wave equation then turns into the system; $\partial_t \Pi - v \Sigma \partial_i \Psi_i = 0$, $\partial_t \Psi_i = v \partial_i \Pi$; You can now choose the unit vector n_i. Note that $P(n_i)$ is already symmetric, H; namely,

the symmetrizer unity and the set hyperbolic. Recall the unit vector is unitary; i.e. $n^2x+n^2y+n^2z=1$, you will find that the eigenvalues of the principal symbol are : $\lambda 1= +v$, $\lambda 2=-v$, $\lambda 3=\lambda 4$; you might have noticed that here we have two degenerate eigenvalues and the system is not strictly hyperbolic; The example above clearly shows that for symmetric hyperbolic sets the eienfields ωa in general depend upon the vector n_i, even when the symmetrizer does not.

The Arnowitt-Deser-Misner equations

Here we rewrite the evolution equations as $\partial t\, u + \partial i\, F(u) = s(u)$, the next step would be to rewrite the Arnowitt-Deser-Misner equations, for the sake of simplicity, the assumption is that you are in vacuum, implying that the matter terms can in any case be considered as sources. The resulting set will be a first order in time/second order in space, as the Ricci; R_{ij} contains second derivatives of the spatial metric γ_{ij}, There are also second derivatives of the lapse in the evolution equation for $\varkappa_{ij}$. In order to have a purely first order set the following quantities should be introduced; $a_i := \partial i\, \ln\alpha$, $d_{ijk} := 1/2\, \partial i\, \gamma_{jk}$; then you will consider that the shift vector β_i is a known function of space and time. The lapse, on the other hand, will be considered a dynamic quantity that evolves through an equation of the form: $\partial t\, \alpha - \beta_i\, \partial i = -\alpha^2\, Q$, with the explicit form of the gauge source function Q to be fixed later; As to the fact that the character structure is only related to the principal part, from now on you can ignore source terms, those that do not contain derivatives of ai, dijk, and Kij. In terms of the dijk, the Ricci tensor can be written as: $R_{ij} \sim -1/2\, \gamma^{lm} \partial_l \partial_m \gamma_{ij} + \gamma_{k(i}\partial_{j)} \Gamma^k$; Now you can write down the evolution equations for ai, dijk and Kij:

$\partial_0 a_i \sim -\alpha\, \partial_i Q$

$\partial_0 d_{ijk} \sim -\alpha\, \partial_i K_{jk}$,
$\partial_0 K_{ij} \sim -\alpha\, \partial_k \Lambda^k{}_{ij}$

$\partial_0 K_{ij} \sim -\alpha\, \partial_k \Lambda^k{}_{ij}$ where $\partial_0 := \partial_t - \beta^k \partial_k$, and where you have been given the definition: $\Lambda^k{}_{ij} :=$ $d^k{}_{ij} + \delta^k{}_{(i}(\,a_{j)} + d_{j)m}{}^m - 2 d^m{}_{mj)}\,)$;
Now you have a set of twenty seven equations to study corresponding to the three components of ai, the eighteen independent components of dijk, and the six independent components of Kij. To proceed with the character analysis you will have to choose a specific direction x and ignore derivatives along the other directions; in that effect you will be analyzing the matrix Mx, instead of analysing the ni Mi; because the tensor structure of the equations makes all spatial directions have precisely the same infrastructure, so it would be enough to analyze one of them. The argument is

then to find twenty seven independent eigenfunctions that will allow you to recover the twenty seven original quantities, where by eigenfunctions here I mean linear combinations of the original quantities u= (ai, dijk, Kij) of the form ωa= ΣCab ub, that up to principal part evolve as ∂t ωa ~ - λa ∂x ωa; with λa the corresponding eigenspeeds. Note that even if the coefficients Cab should not depend on the u's, they can in fact be functions of the lapse α and the special metric γij. By taking into consideration just the derivatives along the x direction; the immediate point will be that there is a set of fields which propagate along the time lines. These fields include: aq, dqij; that gives you fourteen out of likely twenty seven character fields.

The Bona-Masso and NOR formulations

By far you have found provided that the momentum constraints can be guaranteed to be identically satisfied and either the densitized lapse is assumed to be a known function of spacetime, or a slicing condition of the Bona-Masso family can be used, then the Arnowitt-Deser-Misner set would be strongly hyperbolic. Taking a slicing condition of the Bona-Masso type is simple enough, but guaranteeing that the momentum constraints are satisfied in an identical fashion is a different matter; this somewhat point you to the fact that the Arnowitt-Deser-Misner evolution set is a weak hyperbolic and therefore is not well-posed. It is no wonder then that the first fully three-dimensional codes written in the early and mid 90s based on the Arnowitt-Deser-Misner equations had serious stability problems. Well-posed variations of the Einstein equations have in fact been known since the 1950s and were used to prove the first theorems on the existence and uniqueness of solutions, but they were not based on a 3+1 decomposition and required the use of fully harmonic spacetime coordinates, so they were by and large ignored by the numerical community. Efforts to find well-posed versions of the 3+1 evolution equations had to wait until the late 1980s and 90s. Here I am going to concentrate first on the Bona-Masso formulation, which is arguably the simplest way to write down a strongly hyperbolic reformulation of the 3+1 evolution equations. The Bona-Masso formulation starts by defining the three auxiliary variables; $V_i := d_{im} - d_{mi}$.

In terms of the V_i, the Ricci tensor can be written up to principal part as; $R_{ij} \sim -\partial_m d_{ij} - \partial_i (V_j - 1/2 d_{jm}) - \partial_j (V_i - 1/2 d_{im})$. You will then write the evolution equations for a_i, d_{ijk} and K_{ij}, the same as before. If the V_i are independent quantities you will need an evolution equation for them, which can be obtained directly from their definition. In order to write these evolution equations you first consider that, $V^i = 1/2 \, (d^m - \Gamma^i)$, with $\Gamma^i := \gamma^{lm} \Gamma^i_{lm}$ the contracted Christoffel symbols; where the Lie derivative of Γ^i is to be understood as that of an ordinary vector. The fact that the Γ^i are not components of a true vector is what gives rise to the term with the flat Laplacian of the shift. An interesting observation at this point is that you can rewrite the flat Laplacian of the shift as; $\gamma^{lm} \partial_l \partial_m \beta^i = D^2 \beta^i - \mathcal{L}_\beta \Gamma^i - 2\Gamma^i D\beta + R^m{}_m \beta^m$, so that the evolution equation for Γ^i can be written in a more covariant-looking; thereof, the V^i evolve up to pricipal part as: $\partial_0 V^i \sim \alpha(\partial_j K^{ij} - \partial^i K)$; which is a very beautiful result, as it shows that the principal part of the evolution equation for V^i is precisely the same as the principal of the momentum constraints.

Now consider that you have the possibility to add $2\alpha M^i$ to the evolution equation for Γ^i, where $M^i := D_j(K^{ij} - \gamma^{ij} K) = 8\pi j$ are the momentum constraints. You will be of course free to do this as the physical solutions remains the same; you will then have;

$\partial_t \Gamma^i = \mathcal{L}_\beta \Gamma^i + \gamma^{lm} \partial_l \partial_m \beta^i - \alpha \gamma^{il} \partial_l K - \alpha a^l (2 K^i{}_l - \gamma^i{}_l K) + 2\alpha K^{lm} \Gamma^i - 16\pi j$, with $a_i = \partial_i \ln \alpha$. You might want to repeat the analysis you have did before for Arnowitt-Deser-Misner, though now for 30 independent quantities $u = (a_i, d_{ijk}, K_{ij}, V_i)$; this can be done as before by only considering the derivatives along the x direction, you will immediately see that there are now 17 fields that propagate along the time lines, i.e. aq, dqij, Vi.

The Kidder-Scheel-Teukolsky family

The main argument behind Bona-Masso, NOR and BSSNOK for that matter is to introduce three new independent quantities related to contractions of the Christoffel symbols associated with either the physical or the conformal metric, and then to modify the evolution equations for these quantities using the momentum constraints. These formulations also remain second order in space, though they can be written as first order formulation by introducing the first derivatives of te spatial metric dijk as independent quantities and considering their definitions dijk := ∂i γjk / 2 as new constraints. An important point is that in these first order versions the evolution equations for the dijk are obtained directly from their definition and are left unmodified. This has the consequence that the derivative constraints remain trivially satisfied if they were satisfied at the initial point. There is a different approach that can be used to obtain hyperbolic formulation of the 3+1 evolution equations based on the idea of going to a fully first order system and modifying directly the evolution equations for the dijk using the momentum constraints, but without introducing extra independent quantities like the Vi of Bona-

Masso. This approach can be traced back to the most general version of these type of formulations which is known as the Kidder-Scheel-Teukolsky family and has 12 free parameters. This family of evolution equation begins with the derivatives of the spatial metric as independent quantities; $d_{ijk} := 1/2\, \partial_i \gamma_{jk}$, the factor $1/2$ above in fact does not appear in the original Kidder-Scheel-Teukolsky formulation, but here I introduce it in order to have a notation consistent with that of the previous sections; For the Ricci you use again; $R_{ij} \sim -\partial_m d_{ij} - \partial_i (V_j - 1/2\, d_{jm}) - \partial_j (V_i - 1/2\, d_{im})$, with the V_i given by $V_i := d_{im} - d_{mi}$. Note that these are the same quantities used in the Bona-Masso formulation, but here you will not promote them to independent variables and will instead consider them a shorthand for the given combination of d's. It is important to mention the fact that in the Kidder-Scheel-Teukolsky formulation a parameter is added to the above expression for the Ricci tensor R_{ij} to allow for different ways to resolve the ordering ambiguity of the second derivatives of the metric. Now modify the evolution equations for the extrinsic curvature K_{ij} and the metric derivatives d_{ijk} by adding multiples of the constraints in the following manner, $\partial_t d_{ijk} = \alpha \xi \gamma_{i(j} M_{k)} + \alpha \chi \gamma_{jk} M_i$;

$\partial_t K_{ij} = \alpha \eta \gamma_{ij} H$;

with (ξ,χ,η) constant parameters, and where that denotes the original Arnowitt-Deser-misner for the right hand side of the equation prior to adding multiples of the constraints, and where H and Mi are the Hamiltonian and momentum constraints, which up to principal part will be:

$H \sim R \sim -2\partial_m V^m$

$M_i \sim \partial_m K^m{}_i - \partial_i K$

For the analysis of this system you can again use a Bona-Masso type slicing condition $\partial_t \alpha = -\alpha^2 f K$, and introduce the derivatives of the lapse $a_i := \partial_i \ln \alpha$. There is however, an important point related to the slicing condition that should be mentioned. As is done in the NOR set, the Kidder-Scheel-Teukolsky uses a fixed densitized lapse of the form $Q = \alpha \gamma^{-\sigma/2}$ instead of the Bona-Masso slicing condition you will use here.

By densitizing the lapse you are transforming derivatives of a_i into derivatives of d_{im}. Now, as long as the evolution equations for the d's are not modified you find that d_{im} evolves only through K, hence using a Bona-Masso slicing condition is equivalent of densitizing the lapse; however, if the evolution equations for the d's are modified using the momentum constraints.

The null coordinate $\bar{x}^0 := \bar{\tau}_w$ labels null cones whose vertices are along the time-like geodesic $c(\tau)$, where $-\tau_w$ is equal to the proper time

τ along the geodesic, i.e. $\tau w|c(\tau)=\tau$. For points not on $c(\tau)$, $\bar\tau$ is constant along parametrized null geodesics that emanate from $c(\tau)$. We introduce the three spatial coordinates ($\bar x_1, \bar x_2, \bar x_3$). The radial coordinate $\bar x_1 := \ell$ is a positive affine parameter for null rays emanating from $c(\tau)$, where $\ell = 0$. The two coordinates $\bar x_A := (\bar x_2, \bar x_3)$, which are constant both along the time-like geodesic $c(\tau)$ and the null geodesics, label the direction angles of the null rays. In the following, we will use the notation O when we refer to an arbitrary point on $c(\tau)$ that is also the vertex of an arbitrary null cone. For ℓ to be an affine parameter, it has to obey the condition $k^a \nabla_a \ell = 1$, (5) where ∇_a is the covariant derivative with respect to the metric in Fermi normal coordinates. Equation (5) holds, if we define ℓ as in (3) and choose the null vector $k^a(y^b) := (1, y^i/\ell)$ in the Fermi normal coordinate system. Then ℓ does not only measure the space-like distance from the geodesic $c(\tau)$ in a Fermi frame, but it also gives the positive affine distance between a vertex O and points on the respective null cone of O. We note that an affine parameter along a curve is generally defined only up to the transformation $\lambda = a\ell + b$, where a and b are arbitrary constants. However, our choice $\ell = 0$ on $c(\tau)$ implies $b = 0$. From $a = \text{sign}(a)|a|$ and the definition of ℓ (3) one sees that multiplying the Fermi normal coordinates y^i by $|a|$ corresponds to scaling these coordinates, i.e. without loss of generality we can set $|a| = 1$. Hence, null rays $k(\lambda)$ emanating from $c(\tau)$ are most generally parametrized with the affine parameter $\lambda = w\ell$, where $w = \pm 1$. We choose $w = 1$ for future-pointing null rays along $c(\tau)$ and $w = -1$ for past-pointing ones by imposing the normalization condition

$$\lim c_a k^a = -w, \text{ where } \ell \to 0 \quad (6)$$

along $c(\tau)$, where c^α is the tangent vector of $c(\tau)$. A null geodesic $k(\lambda)$ in the null-cone $\bar{\tau}w = \text{const}$ emanating in $\bar{x}^A$-direction, recognizes an expansion in Fermi normal coordinates y^a with respect to the affine parameter λ of the form (using $\lambda = w\ell$ and $w^2 = 1$) where the coefficient functions are evaluated along $c(\tau)$. Setting $\ell = 0$ implies that it describes the timelike geodesic $c(\tau)$ with the tangent vector $c^\alpha = \delta^\alpha{}_\tau$, and that the tangent vector of the null geodesics $k^a = (1/w) dy^a/d\ell |c(\tau)$ does not depend on $\bar{\tau}w$, because the $\bar{x}^A$ are constant along $c(\tau)$. If $c(\tau)$ were no geodesic, an additional dependence of $\bar{\tau}w$ through the directional vector k^a must be taken into account (Newman and Posadas, 1969). Hereafter, we use the null vector

k_a in the form $k_a = 1, n_i(\bar{x}^A)$, where n_i is a three-dimensional unit vector being parametrized by the angles $\bar{x}^A$.

The geodesic equations of the null rays $k(w, \ell)$ read in Fermi normal coordinates

$$d^2 y^a/d\ell^2 = -\Gamma^a{}_{bc}(y^d)\, dy^b/d\ell\, dy^c/d\ell, \quad (8)$$

and further differentiation with respect to ℓ gives

$$d^3 y^a/d\ell^3 = -\Gamma^a{}_{bc,d}(y^g)\, dy^b/d\ell\, dy^c/d\ell\, dy^d/d\ell. \quad (9)$$

Inserting (8) into (9), and equating the thus obtained relations along $c(\tau)$ by setting $\ell = 0$ leads to

$$F^a(\bar{\tau}w, \bar{x}^A) = 0,$$
$$G^a(\bar{\tau}w, \bar{x}^A) = -\Gamma^a{}_{bc,d}|c(\tau)\,(\bar{\tau}w) k^b(\bar{x}^A)\, k^c(\bar{x}^A)\, k^d(\bar{x}^A).$$

References:
[1] M. Alcubierre: Introduction to 3+1 numerical relativity. Phys. 2012;
[2] M. Alcubierre: The appearance of coordinate shocks in hyperbolic formulations of general relativity. Phys.1997
[3] Luisa T. Buchman, Harald P. Pfeiffer, and James M. Bardeen. Black hole initial data on hyperboloidal slices.Phys. Rev.D, 80:084024–1–084024–17, 2009.
[4] J. Frauendiener. Conformal infinity.Living Rev. Rel., 7(1), 2004.
[5] Jeffrey Winicour. Characteristic evolution and matching.Living Rev. Rel., 12(3), 2009.
[6] S. Husa, C. Schneemann, T. Vogel, and A. Zenginoğlu. Hyperboloidal data and evolution.AIP Conf. Proc., 841:306–313, 2006. 28th Spanish Relativity Meeting (ERE05): A Century ofRelativity Physics, Oviedo, Asturias, Spain, 6–10 Sep 2005
[7] V. Moncrief and O. Rinne. Regularity of the Einstein equations at future null infinity.Class. Quantum Grav., 26:125010,2009
[8] M. Chaichian, S. Nojiri, S. D. Odintsov, M. Oksanen and A. Tureanu, "ModifiedF(R) Hoˇrava-Lifshitz gravity: a way to accelerating FRW cosmology",Class. Quant.Grav.27, 185021 (2010), arXiv:1001.4102 [hep-th].[2] S. Carloni. M. Chaichian, S. Nojiri, S. D. Odintsov, M. Oksanen

and A. Tureanu,"Modified first-order Hoˇrava-Lifshitz gravity: Hamiltonian analysis of the generaltheory and accelerating FRW cosmology in power-law F(R) model", Phys. Rev. D82, 065020 (2010), arXiv:1003.3925 [hep-th].

[9] Oliver Rinne. An axisymmetric evolution code for the Einstein equations on hyperboloidal slices.Class. Quantum Grav. ,27:035014, 2010.

[10] Anil Zenginoˇglu and Lawrence E. Kidder. Hyperboloidalevolution of test fields in three spatial dimensions.Phys. Rev. D ,81:124010, 2010.

[11] James M. Bardeen, Olivier Sarbach, and Luisa T. Buchman.Tetrad formalism for numerical relativity on conformally compactified constant mean curvature hypersurfaces.Phys. Rev. D, 83:104045, 2011.

[12] Lars Andersson and Piotr T. Chru′sciel. On 'hyperboloidal' cauchy data for vacuum Einstein equations and obstructi onsto smoothness of scri.Commun. Math. Phys., 161:533–568, 1994.

[13] Jeffrey M. Bowen and James W. York, Jr. Time-asymmetric initial data for black holes and black-hole collisions. Phys.Rev. D, 21(8):2047–2056, 1980.

[14] P.T. Chru′sciel, M.A.H. MacCallum, and D.B. Singleton

. Gravitational waves in general relativity XIV. Bondi expansionsand the polyhomogeneity of scri.Phil. Trans. Roy. Soc. Lond., A350:113–141, 1995

[15] H. P. Pfeiffer, L. E. Kidder, M. A. Scheel, and S. A. Teukolsky. A multidomain spectral method for solving elliptic
equations.Comput. Phys. Commun., 152:253–273, 2003.

[16] R. Arnowitt, S. Deser, and Charles W. Misner. The dynamics of general relativity. In L. Witten, editor,
Gravitation: AnIntroduction to Current Research. Wiley, New York, 1962.

[17] H. Bondi, M. G. J. van der Burg, and A. W. K. Metzner. Gravitational waves in general relativity VII. Waves from
axi-symmetric isolated systems.Proc. R. Soc. Lond. A, 269:21–52, 1962.

[18] R. K. Sachs. Gravitational waves in general relativity. VIII. waves in asymptotically flat space-time.Proc. R. Soc. Lond.
A, 270(1340):103–126, October 1962.

[19] L´aszl´o B. Szabados. Quasi-local energy-momentum and angular momentum in General Relativity: A review article.
LivingRev. Rel., 12(4), 2009.

[20] Piotr T. Chrusciel, Jacek Jezierski, and Szymon Leski. The Trautman-Bondi mass of hyperboloidal initial data sets.Adv.
Theor. Math. Phys., 8(1):83–139, 2004.
[21] Roger Penrose. Asymptotic properties of fields and space-times.Phys. Rev. Lett., 10(2):66–68, 1963.
[22] J. M. Stewart. Numerical Relativity III. the Bondi massrevisited.Proc. R. Soc. Lond. A, 424:211–222, 1989.
[23] P.T. Chrusciel, M.A.H. MacCallum, and D.B. Singleton. Gravitational waves in general relativity XIV. Bondi expansions
and the polyhomogeneity of scri.Phil. Trans. Roy. Soc. Lond., A350:113–141, 1995.
[24] R. Penrose. Conformal treatment of infinity. InRelativity, Groups, and Topology, pages 565–584. Gordon and Breach,
New York, 1964.
[25] Louis A. Tamburino and Jeffrey H. Winicour. Gravitational fields in finite and conformal Bondi frames.
Phys. Rev.,150:1039–1053, 1966.
[26] Piotr T. Chrusciel, Jacek Jezierski, and Malcolm A. H.MacCallum. Uniqueness of the Trautman-Bondi mass.Phys. Rev.
D, 58:084001, 1998.

[27] Demetrios Christodoulou. Reversible and irreversible transformations in black-hole physics.Phys. Rev. Lett., 25(22):1596–1597, Nov 1970.

[28] Gregory B. Cook and James W. York, Jr. Apparent horizonsfor boosted or spinning black holes.Phys. Rev. D, 41(4):1077–1085, 1990.

[29] Helmut Friedrich. Cauchy problems for the conformal vacuum field equations in general relativity.Commun. Math. Phys.,91(4):445–472, 1983.

[30] Helmut Friedrich. On the existence of n-geodesically complete or future complete solutions of Einstein's field equations with smooth asymptotic structure.Commun. Math. Phys., 107(4):587–609, 1986.

[31] Gregory B. Cook. Corotating and irrotational binary black holes in quasicircular orbits.Phys. Rev. D, 65(8):084003, Mar2002.

[32] Philippe Grandcĺement, Eric Gourgoulhon, and Silvano Bonazzola. Binary black holes in circular orbits. II. Numericalmethods and first results.Phys. Rev. D, 65:044021, 2002.

[33] R. L. Arnowitt, S. Deser and C. W. Misner, "The dynamics of general relativity",arXiv:grc/0405109, originally inGravitation: An Introduction to Current

Research,L. Witten ed., John Wiley & Sons Inc., New York, London, 1962. Republished in Gen.Relativ. Gravit.40, 1997[] James W. York. Conformal "thin-sandwich" data for the initial-value problem of general relativity.Phys. Rev. Lett.,82(7):1350–1353, Feb 1999

[34] James M. Bardeen and Luisa T. Buchman; Physics Department, University of Washington, Seattle, Washington 98195 USA and Theoretical Astrophysics, California Institute of Technology, Pasadena, California 91125 USA: March 19, 2012

www.ingramcontent.com/pod-product-compliance
Lightning Source LLC
Chambersburg PA
CBHW070333190526
45169CB00005B/1869